Weather Mania

DISCOVERING WHAT'S UP AND WHAT'S COMING DOWN

Michael A. DiSpezio

Sterling Publishing Co., Inc.
New York

Photo credit: Page 80, author photo by Anthony DiSpezio

Acknowledgment

Once again, I've had the good fortune of working with my Sterling editor, Hazel Chan. Hazel and I have worked together on a variety of projects, including this new Mania series. It's Hazel's editorial skills that account for the success of these works. I'd also like to thank the book's artist, Dave Garbot for his wonderful, humorous, and kid-friendly approach to illustrating. Finally, I'd like to thank Sheila Anne Barry for supporting and helping guide the Mania series.

Design by Nancy B. Field

Library of Congress Cataloging-in-Publication Data Available

10 9 8 7 6 5 4 3 2 1

Published by Sterling Publishing Co., Inc.
387 Park Avenue South, New York, N.Y. 10016

© 2002 by Michael A. DiSpezio

Distributed in Canada by Sterling Publishing
c/o Canadian Manda Group, One Atlantic Avenue, Suite 105
Toronto, Ontario, Canada M6K 3E7

Distributed in Great Britain and Europe by Chris Lloyd at Orca Book
Services, Stanley House, Fleets Lane, Poole BH15 3AJ, England.

Distributed in Australia by Capricorn Link (Australia) Pty. Ltd.
P.O. Box 704, Windsor, NSW 2756 Australia

Printed in China

Sterling ISBN 0-8069-7745-0

Contents

Weather Wise?

You are in charge of scheduling the weekend's ultimate outdoor event. Before you lock in the date, however, you need to check on one thing. If you're like most people, that one thing will be the weather.

You switch on the radio just in time to hear the end of the weather report: "Low pressure system moving into the region on Saturday night."

You then hit the remote for the TV and catch the end of the weatherperson saying, ". . . and that's the weekend outlook. Plenty of cumulonimbus on Sunday. And now, sports."

You pick up a paper only to discover that your younger sister has drooled all over the weather page. Yuck. The only part that's slobber-free is the weather map. At least it's a start.

You begin reading the map, but you have no idea what the symbols mean. And even if you could tell a millibar from a minivan, you still wouldn't have a clue to what was going on.

What do you do? Feel as low as a tropical depression? Hardly, especially since you have in your hands the ultimate kids' book on weather, *Weather Mania*!

So take a breather, sit back, relax—and DO this book. Flip through the pages. You'll find yourself in a learning adventure that's packed with all sorts of things about the weather, including cool facts, experiments, weather sayings, and bad jokes.

Get Above It

"This is your captain speaking. We have some bad weather ahead of us. We'll try to get above it. But until then, please return to your seats and secure your seat belts."

Get above bad weather? What gives? Is this some sort of spiritual challenge?

Actually, her announcement is a lot more "down to Earth" than you may realize. You see, weather only occurs in the lowest layer of our planet's atmosphere. This layer of air, called the *troposphere* (TROH-puh-sveer), stretches skyward to a height of about 7 miles (11 kilometers). This upper boundary isn't fixed. Sometimes the air pushes the top of the troposphere up to 10 miles (16 km).

At the upper limits of the troposphere and beyond, the air is mostly still and unchanging. So if a craft encounters turbulence, then climbing to a higher altitude (height) may place it above the weather's roller-coaster ride.

WHAT'S WEATHER?

Weather describes the everyday changes that occur in our surrounding layer of air. Rain, snow, sleet, drought, winds, hurricanes, storms, tornadoes, blizzards, hail, heat waves, frosts, and sleet are some of the changes that we think of as weather. Even what you might consider as no weather is weather. A dry, still, cloudless day is a change from a previous or future day in which there's more humidity, wind, or clouds.

SOFT EDGE

How high up does our atmosphere go? Some scientists say that it goes skyward to an altitude of 80 miles (129 km). Others place the upper limit at 180 miles (290 km). Others say it's somewhere in between. Okay, so who should you believe? And even more important, what altitude should you write on a science report? The answer is as cloudy as an overcast day. It all depends.

As you go higher into the atmosphere, it gets thinner and thinner and thinner. . . . Eventually, the atmosphere has the same super-low number of gas particles as outer space. You get the picture? Our planet's thinning air gradually becomes outer space.

WELCOME TO SPACE

ATMOSPHERE ENDS HERE

Going Up?

Imagine the hottest day of the summer. Temperatures skyrocket. However, you are one lucky camper. You're flying off on vacation. As you look out your window, you see a spot of frost on the outer plastic. Strange, it was hot outside when you left the ground—and you were headed south! What gives? Okay, think quick.

On a hot summer's day, how cold is the air temperature outside of a jet flying at an altitude of around 25,000 ft (7.6 km)?

a) The same as the temperature of the ground that is directly below the craft.
b) 32°F.
c) −30°F.

Answer: c) A chilling surprise: It's −30°F (−34°C) out there. That's because as you rise up in the troposphere, the temperature keeps dropping. Brrrrrr.

MAJOR SCALES

Fahrenheit Scale: This is the temperature scale used by most people living in the United States. It's based on 32°F for the freezing temperature of water and 212°F for its boiling point.

Celsius Scale: This is the common scale found almost everywhere outside the United States. It's based upon 0°C for the freezing point of water and 100°C for its boiling point.

CONVERTING TEMPERATURE

To convert from Celsius to Fahrenheit:
Multiply the temperature in Celsius by ⁹⁄₅
and then add 32. The official equation goes:

$$°F = \tfrac{9}{5} \times °C + 32.$$

To convert from Fahrenheit to Celsius:
Subtract 32 from the Fahrenheit temperature
and then multiply by ⁵⁄₉. The official equation
goes:

$$°C = (°F - 32) \times \tfrac{5}{9}.$$

SCALING RECORD TEMPERATURES

Lowest temperature recorded on Earth:
−129°F (−89.4°C) on July 21, 1983 in
Vostok, Antarctica

Highest temperature recorded on Earth:
136°F (57.8°C) on Sept. 13, 1922 in
El Azizia, Libya

The Heat Is On

The sun is blazing and you're at the beach. The sand is so hot it feels like the surface of the sun. Although you're wearing protective clothing and covered with gobs of sunscreen, you don't have any shoes. Your tender little soles are bare and can easily burn.

You're standing on a blanket at the X. Your friends are hanging at the Y. What path do you choose to cross the burning sands?

Answer: If you're smart, you won't cross the sand at all. Instead, you'll wade in the shallow water. Although the sand is heated to a foot-scorching temperature, the water stays cool.

BY LAND AND BY SEA

Land and water don't heat up (or cool down) at the same rate. But don't take our word for it. Try the following.

You'll Need
Two plastic foam containers
Water
Soil or sand
Two small outdoor thermometers

To Do
1. Fill one of the plastic foam containers with room temperature water.
2. Fill the other container with soil or sand.
3. Carefully insert a thermometer into the soil. You may have to tilt the container on its side to make room for the thermometer. Record the temperature of the soil.
4. Insert the other thermometer into the water. Record the temperature of the water.
5. Place both cups in direct sunlight (or under a heat lamp).
6. Measure the temperature of each material every 5 minutes. Record this data.
7. After an hour, compare and contrast the difference in temperature change.

MAKING THE CONNECTION

So what does all this hot and cold stuff have to do with weather? Everything! The differences in the way things heat up is the "engine" that drives *all* of the wind and changes in the atmosphere.

Cell, the Concept

When the sun's rays strike the surface of our planet, things begin to heat up. As this surface warms, it transfers heat energy to the surrounding air. The air absorbs this energy and warms up.

When things heat up, they spread out (expand). Air is no exception. The warmed air expands and its particles become less crowded together (less dense). This less-dense air begins to rise.

As air rises, it cools. The cooler it becomes, the more it draws together (contracts). The more it contracts, the denser it gets. As its density increases, air becomes heavier. Eventually, the air becomes denser than the surrounding air. At this point, the cooled-off air begins to sink.

Rising and sinking air can't exist in the same place. As it travels up and down, the movement squeezes, pulls, and pushes air out of its way. These sideways air movements create wind!

CONVECTION CELL

This cycle of rising, sinking, and horizontal air movements is called a *convection cell*. It is called a cell because it is a closed loop.

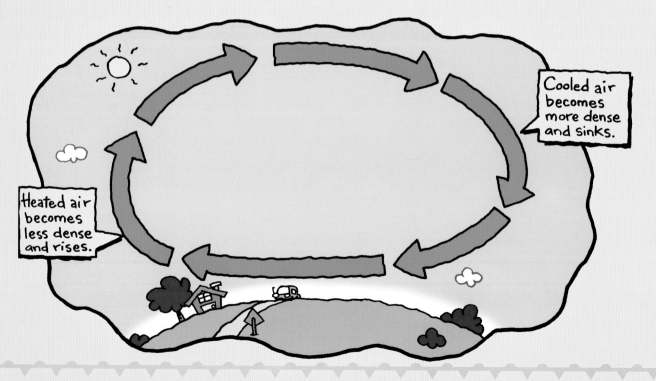

Cooled air becomes more dense and sinks.

Heated air becomes less dense and rises.

AIR SERPENTS

The rising of warmed air occurs all over the place, even in your bedroom. Here's a neat little device called an air serpent that spins when placed in a stream of moving air.

You'll Need

Paper	Tape
Pen	Thread
Scissors	

To Do

1. Trace this serpent drawing onto a piece of paper.
2. Use scissors to carefully cut out the pattern of the coil so that the serpent hangs freely.
3. Use tape to attach a length of thread to the serpent's head. Attach the other end of the thread to the ceiling or the top of a doorway.
4. Watch the serpent spin slowly in moving air.

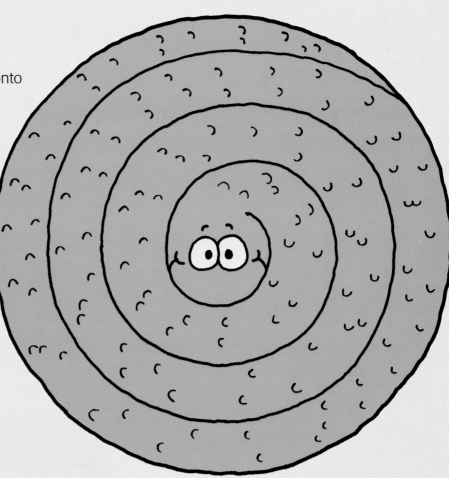

Up, Down, and Curved?

At the equator, air heats up and rises. Imagine that this warm air moves in the upper troposphere towards the North or South Poles. As this air gets to either of these poles, it cools and sinks. When it reaches the land surface of the North or South Pole, the cooled air then travels back to the equator. Nice and simple—too bad that's not how it works out.

The big cell separates into smaller cells. Three are north of the equator and three are south of the equator. The movement of air along the bottom of these six cells forms areas called wind belts.

According to our scheme, winds should blow due north and due south. They don't! There's a catch—and that catch is called the Coriolis (CORE-ee-oh-lis) force.

CORIOLIS FORCE

The Earth's spin produces something called the Coriolis force. This force bends the paths of all moving objects. In the Northern Hemisphere, all paths curve to the right (clockwise). South of the equator, the bends are to the left.

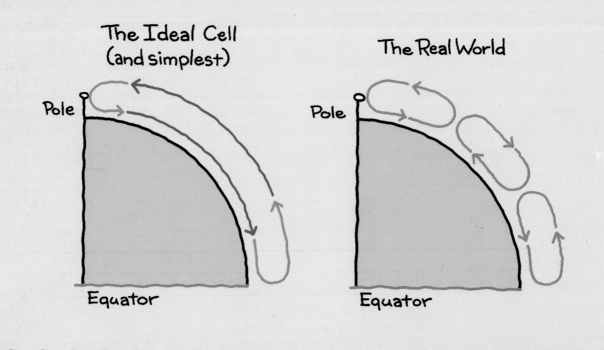

The Ideal Cell (and simplest)

Pole

Equator

The Real World

Pole

Equator

Without the curving of paths from the Coriolis force, the winds would blow mostly along straight paths.

Without bending from the Coriolis force

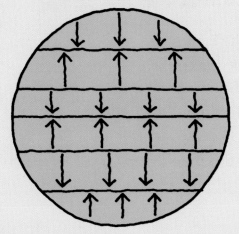

With bending from the Coriolis force

THE BATHROOM TEST

Some scientists say that the Coriolis force can even be seen in sinks, bathtubs, and toilet bowls. As the water flows down the drain, it spirals in either a clockwise or counterclockwise direction. Other scientists say that the scale within a kitchen sink or basin is too small to make any difference.

Blowing in the Wind

A north wind blows from which direction?

a) North.
b) South.
c) Over there.

Answer: a) North. Winds and weather are named after the direction from which they come. A northern wind is a wind that blows in from the north. Likewise, a southern gale is a storm that comes in from the south.

BLOWING BUBBLES

You can uncover the directions of breezes and local winds with bubbles. Blow a slow steady stream of bubbles and observe their path. Since bubbles don't weigh much, even the lightest wind will carry them about.

WIND SOCK

Have you ever seen a big orange sock on an airport runway? If so, you've seen a weather instrument that tells the pilot which way the wind is blowing. The large size and bright color makes this weather instrument visible from great distances (as in looking down to the ground from the sky).

Here's how to build a smaller version of a wind sock—without the airport runway.

You'll Need

Bendable wire (found in
craft stores; 1½ feet/about ½ m)
Coffee can
Nylon stocking (such as tights,
but not pantyhose)
or lightweight fabric (such as the material
used for assembling kites)
Scissors
Tape
Short stick (1–2 feet/about ⅔ m)
Kite string (1 foot/about ⅓ m)
Pushpin

To Do

1. Shape the wire into a circle that has the diameter of a coffee can.
2. Cut one of the legs of the stocking so that you have the thigh to the ankle.
3. Tape the thigh-end of the stocking to the wire. The cut-off end remains unattached.
4. Insert a short stick in the ground.
5. Attach a 6-inch (15-cm) length of kite string across the wire circle. The ends of this string should be positioned directly across the wire. Attach another 6-inch (15-cm) segment of string to the center of this string.
6. Secure a pushpin in the top of the short stick. Tie the free end of the string to the pushpin. The stocking assembly should be free to spin in all directions.

Wind Watching

Close your eyes. Imagine being at the beach on a hot summer's day. Not bad, eh? As you sit on your blanket, you feel a cooling breeze come off the ocean. Is this breeze a chance happening or is it science at work?

It's science. The breeze is called a *sea breeze*. Like winds, breezes are also named by the direction from which they come.

The sea breeze comes from the unequal heating of the water and the land. Land heats up first. The air directly above the land rises. The cooler air that was in place over the ocean rushes in to fill this gap. This sideways rush of air is the sea breeze.

At night, the tables are turned. The land cools off quicker than the water. Air above

the water remains warmer than air above the cooler land. This warmer air rises. Air over the land moves out to fill the gap. This sideways rush of air is called a *land breeze*.

WHICH WAY?

Most of the weather that moves across the United States travels:

a) North to south.

b) East to west.

c) West to east.

d) Along route I-95.

Answer: c) West to east. This should sound familiar to anyone who's ever watched a national weather report or looked at several days of national weather maps. Although weather can go in all directions, it mostly follows a west-to-east path. This movement is powered by the mostly west-to-east wind direction found at this latitude.

WIND STREAMER

Like a wind sock, a streamer can also show the direction of the wind.

You'll Need

Marker	Scissors
Paper plate	Tape
Streamer	Compass

To Do

1. Use a marker to write the directions north, south, east, and west on a paper plate.
2. Cut a 1-foot (about ⅓-m) section of paper streamer.
3. Use tape to secure the streamer's end to the center of the plate.
4. Hold the plate parallel to the ground.
5. Using a compass, find north. Turn your plate so that its north points in the same direction. See how the streamer responds to the wind.

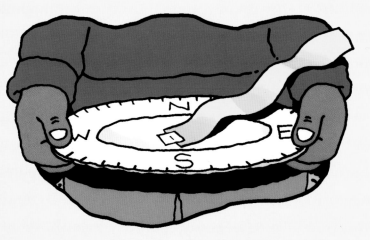

Pointing the Way

Perhaps the most familiar instrument that shows wind direction is the weather vane. Also called a wind vane, this rooftop ornament has some sort of arrow that "points" to the wind's direction.

You'll Need

Heavy stock paper   Scissors
Straw   Glue
Tape   Magnetic compass
Shoebox   Books
Empty thread spool

To Do

1. Cut an arrow shape out of a sheet of heavy stock paper. The arrow needs to have a long slender shaft.
2. Insert the shaft into a straw. Use tape to secure these parts together.
3. Punch a hole in the top of a shoebox.

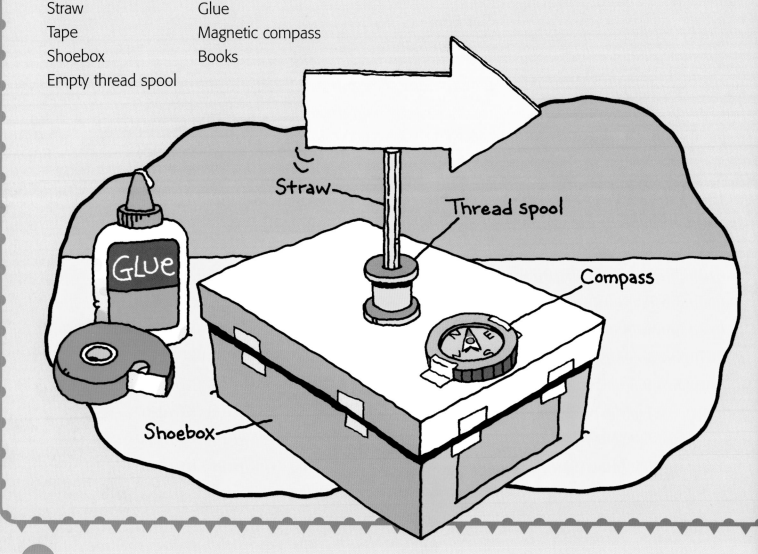

Straw

Thread spool

Compass

Glue

Shoebox

4. Position a spool over the opening so that the holes align. Glue the spool to the box top.
5. Insert the straw through both holes. The straw and arrow should spin freely.
6. Tape a magnetic compass to the top of the box. The compass needle will point out true directions.
7. To prevent the weather vane from blowing away, load weights (such as the textbooks you haven't opened all year) into the box and then secure the shoebox cover with tape.

WEATHER VANE (THE QUICK VERSION)

Use a pushpin to anchor a feather into the eraser of a pencil. Make sure that the hole punched into the quill of the feather is large enough to allow the feather to freely spin.

Historic Connection

During ancient battles, warriors would place flags on the battlefield. These flags flapped in the wind. Archers gathered information about the air currents by observing the flag movements. Taking into account these movements, the archers improved their shots with deadly accuracy.

Measuring Pressure

Do you feel a weight on your shoulders? No, we're not talking about next week's French exam or the homework your dog ruined. The pressure we're talking about is a force pushing down on you. Actually, the force is not only pressing on you, but it's also pushing in on your sides, back, front, arms, and legs. This unseen force is called *air pressure*.

Air pressure is produced by the weight of the air above you. The more air that's on top of you, the greater the air pressure. The less air, the lower the pressure.

Impress an Adult!

Air pressure can be measured in all sorts of units. Meteorologists, however, prefer to report pressure in inches of mercury. This measurement reflects the height at which a column of mercury will be supported by air pressure. At sea level, the average air pressure is 29.92 inches of mercury.

Weather maps often use a different unit. It is called a millibar. At sea level, there is an average pressure of 1013.25 millibars.

And let's not forget divers, engineers, and stateside auto mechanics who prefer to use pounds per square inch as their pressure unit. For them, it's 14.7 lbs/square inch.

Sea Level
1013.25
Millibars

BAROMETER BUILDING

The weather instrument that measures air pressure is called a *barometer*. You can build a simple barometer using a stiff plastic container, rubber balloon, scissors, broom straw (or long toothpick), glue, clay, index card, and marker.

1. Slit the rubber balloon and stretch it across the opening of a rigid plastic container.

2. Place a spot of glue in the center of the stretched balloon. Set the broom straw in the glue so that it becomes attached to this soft surface.

3. Use a small lump of clay to position a pressure scale next to the straw. Copy the gauge of the scale as shown in the illustration.

As the air pressure increases, the force pushes down on the balloon skin. The middle part of the skin "caves" in. This causes the broom straw to angle upwards and point to a higher pressure reading. As the air pressure decreases, the middle part of the balloon skin rises and the broom straw points downward.

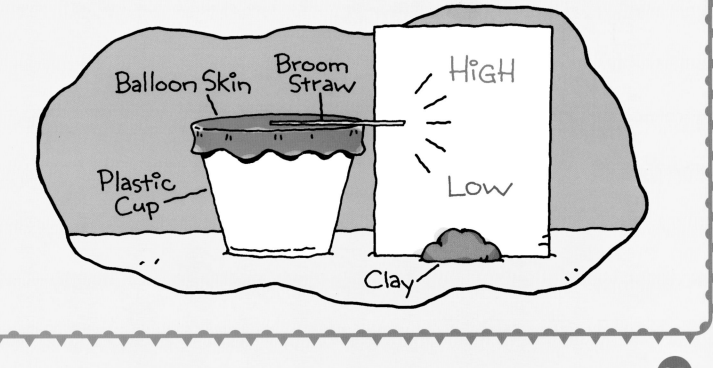

Balloon Skin   Broom Straw   HIGH   Low

Plastic Cup   Clay

23

Pressured into Moving

Imagine a room packed with party-goers. Yahoo! However, the crowd is so large that there's hardly any space to stand. Squished together, the crowd begins to get edgy. Everyone wants more space. Everyone wants out.

Now imagine a room next door that's the same size as the crowded one. Just a few people share the huge space. With room to spare, there's no pushing, shoving, or pressure to leave.

Suddenly, a large doorway is opened between the two rooms. What happens? It doesn't take high-level thinking to imagine people rushing from the crowded room into the empty one. This forceful movement spreads out the crowd. However, once the numbers are equal, there's no more tension to move between the rooms.

Although we've been talking about people in a party, the same holds true for the atmosphere. Think of a space that is

packed with a lot of air particles. We'll call this a *high pressure* region. As the number of air particles increases, the force to spread out also increases.

Now imagine a space next to it with very few particles. We'll call this a *low pressure* region. Like the room-next-door situation, air from the high pressure region will flow into the low pressure area. It's this movement of air that we call wind.

Blown Away

Match the place with its maximum wind speed. Then learn more about your answers below.

Earth	**0 mph (0 km/hr)**
Moon	**231 mph (372 km/hr)**
Jupiter	**330 mph (531 km/hr)**

Moon - 0. There's no wind on the moon. Zero. Zilch. Zip. That's because there's no air. Actually, there is a tiny trace of air, but it's so small that it doesn't really amount to anything. So although the moon has awesome temperature differences, there is no atmosphere that can be whipped up by the sunlight that strikes its surface.

Earth - 231 mph. On August 12, 1934, the highest wind gusts ever recorded on the surface of Earth were measured on top of Mt. Washington. For a 2-day period, these winds averaged nearly 130 mph (209 km/hr)!

Jupiter - 330 mph. Jupiter's atmosphere is made up mostly of ammonia, ice, and weird crystals that form a large swirling slosh that is visible in telescope images. Within these bands of slosh are huge thunderstorms that would be large enough to cover two-thirds of the United States. About 300 years ago, a giant red spot appeared in the atmosphere. Astronomers believe that the spot is a superstorm that's nearly twice as wide as planet Earth!

JET STREAM

The jet stream is a fast wind that is found at about 6–7 miles (10.5 km) in altitude. For the record, it is defined as a fast-moving current of air (over 57 mph [92 km/hr]) that divides the warm air mass of one region from the cooler, more northern air mass of another region. The strongest jet streams can reach speeds of 190 mph (306 km/hr)!

When the jet stream dips south, it carries cold Canadian air into the midsection of the United States. Temperatures drop. If the jet stream maintains a more northern track, warmer air moves north from more southerly regions.

Cold Air Mass

L

H

Beaufort Scale

What is the Beaufort Scale?

a) Rust deposits in pipes and plumbing.
b) Musical notes common to North Carolina music.
c) A way to observe, measure, and compare wind speeds.

Answer: c) A way to observe, measure, and compare wind speeds. The scale was created in 1805 by British Admiral Sir Frances Beaufort. The scale was originally used to help sailors. Eventually, it was expanded to include land-based observations.

BEAUFORT'S BEST

According to Beaufort, the air was calm (0–1 mph [0–1.6 km/hr]) if the sea resembled the surface of a mirror. Ripples were produced by "light air" (1–3 mph [1.6–5 km/hr]). Next came an assortment of increasing breezes with wind speeds from 3 to 31 mph (5–50 km/hr). Then came stronger winds that Beaufort called gales (32–54 mph [52–87 km/hr]). Next came storms. These dangerous winds produced high waves, foam, and limited visibility. When a storm wind reached 74 mph (119 km/hr), it was raised to hurricane status.

Beaufort Scale

0-1 MPH = Calm
1-3 MPH = Light
3-31 MPH = Strong
32-54 MPH = Gales
74+ MPH = HURRICANE (Yikes!!)

OF WINDS AND WAVES

Wind that goes along the sea surface produces tiny ripples. As the wind continues, these ripples grow into waves. Under the right conditions, the waves can grow to enormous heights!

THE PERFECT WAVE

During Halloween, 1991, three different weather systems bumped into one another off the coast of Nova Scotia. This awesome weather event produced a storm of gigantic proportions. Its weather stretched up and down the length of the east coast of North America.

Winds that were over 100 mph (161 km/hr) blew over a large stretch of open water. This churned up the sea into towering waves. Meteorologists estimated the height of the largest waves at 150 feet (46 m). They were as tall as a 15-story building!

NOVA SCOTIA

How Fast Is Fast?

The weather watcher's method that best measures wind speed uses:

a) A pinwheel.

b) An anemometer.

c) Homework pages scattered in an unexpected breeze.

Answer: b) An anemometer.

A LITTLE HISTORY

The first anemometer (an-e-MOM-e-ter) was a swinging arm device. The force of the wind pushed against a hanging ball. As the wind speed increased, the ball climbed upwards. Its climb moved an attached pointer along a curved scale. The greater the force, the higher the pointer moved. The lesser the force, the smaller the climb.

You can assemble a modern-day version of this swinging arm anemometer using a protractor, string, table tennis ball, and tape. Simply tape a 6-inch (about 15 cm) length of string to a table tennis ball. Attach the other end of the string to the center of the straight edge of a protractor. As the wind blows, the ball will rise up along the measurement scale.

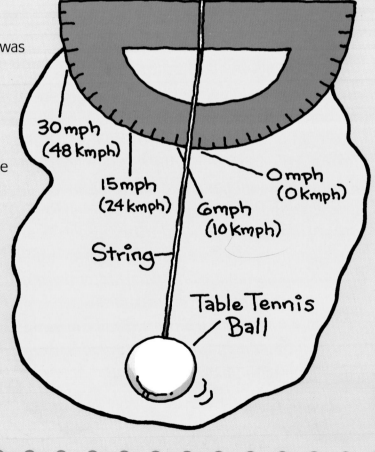

30 mph
(48 Kmph)

15 mph
(24 Kmph)

6 mph
(10 Kmph)

0 mph
(0 Kmph)

String—

Table Tennis
Ball

A ventimeter (ven-ti-ME-ter) is another type of wind-measuring device. As wind blows across the opening of a closed tube in the ventimeter, it produces a pressure drop in the upper end of the tube. Since the pressure still remains high beneath the ball, the difference produces a force that pushes a ball up the tube.

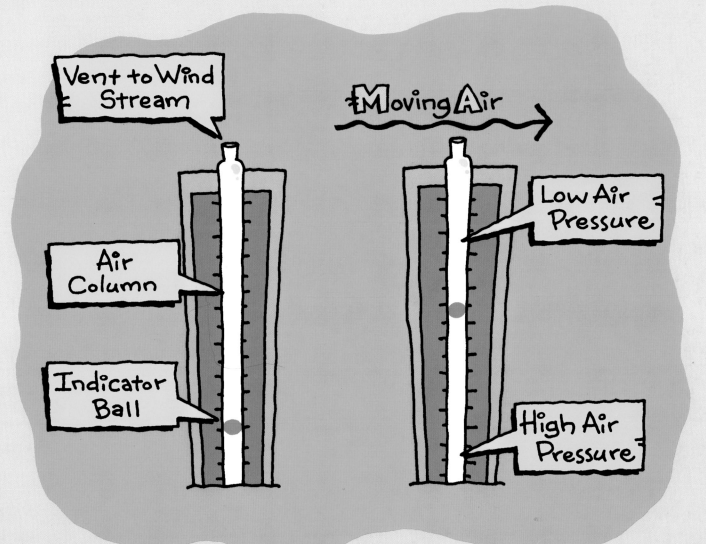

Twirling Cups

A cup anemometer is another type of wind-speed indicator. This tool has a series of cups that are positioned around a rotating axle. Perhaps you've seen this type of tool attached to the roofs of buildings, backyard weather stations, or to ship masts?

The blowing wind is "caught" by the cup of one bowl. In response to the push, the cup rotates. This spin advances the next cup into a position in which it "catches" the wind. As this cup rotates, the next cup spins into position. So on and so on and. . . .

The simplest way to measure the wind speed is by observing the cup movement. The faster the spin, the greater the wind speed. The slower the spin, the lower the wind speed.

BUILDING YOUR OWN CUP ANEMOMETER

You'll Need
Four paper or
   plastic foam cups
Tape
Two wooden craft sticks
Glue
Empty thread spool
Shoebox
Books
Scissors
Plastic straw
Lump of modeling clay

Anemometer

1. Use tape to attach one cup to the end of a craft stick. The cup should "hang" from the stick and face sideways to the straw's length.

2. Attach a second cup in the same manner to the opposite end of the stick. Make sure that the two cups open in the opposite direction.

3. Repeat steps 2 and 3 using another stick and two additional cups.

4. Use glue to secure one side of the spool to the shoebox lid.

5. Fill the inside of the shoebox with books so that it doesn't blow away. Replace the lid and tape it down.

6. Use the scissors to cut the straw in half. Insert a straw half into the hole in the center of the spool. It should spin freely.

7. Place a coin-sized lump of clay over the exposed end of the straw.

8. Overlap the two sticks at their centers to form an "X." Push this overlapped section into the lump of clay so that the sticks are held in place.

9. Place the anemometer in the wind and watch the cups whirl.

GETTING NUMBERS

To get some sort of handle on comparing wind speeds, you'll need to find out how fast the cups are spinning. The easiest way is to paint one of the cups with a bright color. How about a sunny orange? Focus only on this cup and you'll have a much simpler job in measuring the spin rate.

Water, Water, Everywhere

The air is loaded with water. Although you might sense its presence, you can't see it. That's because the water in air isn't a liquid (unless of course it's raining). The water that fills the air is a gas. And that gas is invisible.

WHERE'S THE WATER?

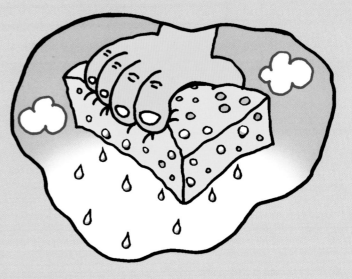

When a droplet of water turns into a gas, each water particle (molecule) scatters off into a different direction. So instead of a big "glob" of water, you have billions and billions and billions of individual water particles spreading out and mixing in among the gas particles in the air. The individual water particles are so tiny that you can't see them.

Although it doesn't look it, think of the atmosphere as a sponge. If you examine it closely, you will find holes in the sponge that offer a space for all sorts of particles, including water molecules. The bigger the holes or the more holes there are, the more water particles will fit into the atmosphere. When the holes begin to shrink, water is forced out of the atmosphere as if the sponge is being squeezed. As the water particles leave the air, they enter the liquid phase. Tiny droplets of liquid come together and grow into larger drops. Eventually, the drops becomes too heavy to remain up in the air. The larger droplets fall as rain or snow.

WATER CYCLE

To better understand the relationship between water and our planet, why not build a model of the water cycle? Put some soil on one side of a plastic tank to represent land. Put some water on the other side to represent the ocean. Place a clear plastic lid on top of the tank. Put a small bowl of ice cubes on the lid to model the colder temperatures of higher altitudes. Set the tank in the sunlight and watch what happens.

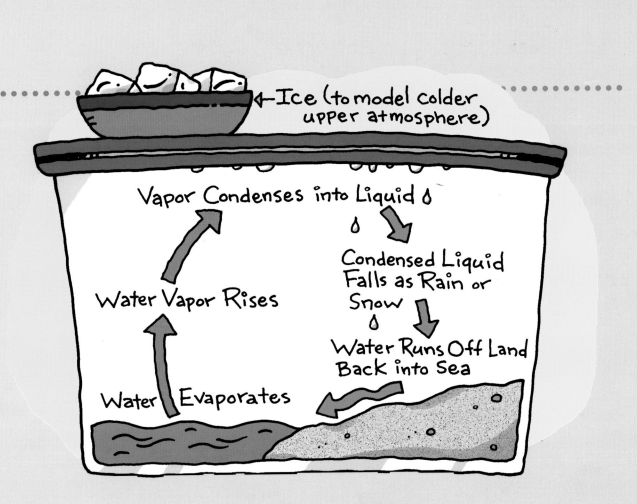

Ice (to model colder upper atmosphere)

Vapor Condenses into Liquid

Water Vapor Rises

Condensed Liquid Falls as Rain or Snow

Water Runs Off Land Back into Sea

Water Evaporates

Impress an Adult!

As water gains heat, its particles speed up. Some get going so fast that they shoot out of the liquid into the surrounding air. This change from liquid to gas is called evaporation (e-va-por-A-shun).

In contrast, when water vapor cools, its particles slow down. When they lose enough energy, the particles of gas change back into liquid. This change is called condensation (con-den-SA-shun).

Water that has condensed in the sky can fall to Earth as rain, snow, sleet, or hail. These forms of weather are called precipitation (pre-sip-i-TA-shun).

Measuring Humidity

Water vapor that's in the air is called *humidity*. It's this humidity that makes the air feel muggy or heavy with moisture. Drop the humidity and air feels dry and crisp.

The temperature of the air determines how much water vapor it can hold. Warmer air can hold more water vapor than cooler air. That's because the particles in warm air move quicker and spread out more. Since these air particles are further apart, there is more space for other particles, such as water molecules.

Relative humidity is a measurement that compares the amount of water vapor actually in the air to the maximum amount of water vapor that the air can hold. If the air is holding only half of its maximum load, the relative humidity is 50%. If the air is at its limit, it's full with 100% humidity.

GET YOUR HANDS WET

A hygrometer (high-GROM-e-ter) is an instrument that detects and measures the amount of water in the air. Here's a type of hygrometer that is based on a balanced arm design.

You'll Need

Index card	Pin
Scissors	Straw
Paper cup	Scraps of blotter paper
Tape	Modeling clay

TODAY'S FORECAST
Warm Temps
90% Humidity

To Do

1. Cut an index card in half. These halves will form the support for the balance arm.
2. Turn a paper cup upside down. Use tape to attach each half of the index card to opposite sides of the cup bottom. As shown, each card half should stick up from the flat cup bottom.
3. Insert a pin through both index card halves and through the center of the straw. The pin should act as an axle, allowing free movement of the straw.
4. Tape several pieces of blotter paper together. Secure this stack of paper to one end of a straw.
5. Place a lump of modeling clay on the opposite end of the straw. Adjust the amount of clay so that the straw balances.
6. Experiment with your hygrometer by moving it to places of different humidity. Try taking it into a bathroom in which the shower has been running for several minutes. What happens to the balance?

Turning Up the Heat

"It's the humidity, not the heat."

No doubt you've heard that expression before. But what's the real story? When it's 100°F, it's hot. Who cares what the humidity is. Right? Wrong. When it's hot, your body sweats to help bring down your temperature. The sweat on your skin's surface absorbs body heat. When the sweat evaporates into the surrounding air, the skin cools off. This type of cooling is called evaporative cooling. It is found everywhere from air conditioners to beehives.

But when the humidity is high, there's less room in the surrounding air to hold your sweat. So although you perspire, the liquid just remains on your skin. Yuck.

In drier air, it's a different (and much prettier) story. It begins the same. You heat up and perspire. However, since the air is dry it can hold your evaporated sweat. The evaporating sweat absorbs heat, removing it from your skin. Since you lose heat, you feel the immediate relief of evaporative cooling.

APPARENT TEMPERATURE

The National Weather Service has a chart that shows how the temperature and humidity work together to create an apparent temperature—what the temperature feels like. The actual temperature is what the number on the thermometer says. Here's a part of that chart:

Temperature (°F)	Relative humidity (%)				
	0	25	50	75	100
110	99	117	150	Too hot to handle	
100	91	101	120		
90	83	88	96	109	
80	73	77	81	86	91
70	64	66	69	70	72

From this chart you can see that a day on which the actual temperature is 110°F, the apparent temperature can change as much as 50 degrees. If there's 0% relative humidity, then sweating works great. So although it's 110°F, it will feel like a balmy 99°F. If, however, the relative humidity is at 50%, you'll be swimming in hot sweat. Not only will you be drenched, but the apparent temperature will be a dangerous 150°F!

Cool Fact

Dogs don't sweat. They do, however, use evaporative cooling to beat the heat. Instead of sweat, they use spit (how nice). As a dog pants, it exposes its tongue to the surrounding air. Like sweat, saliva absorbs heat from the exposed moist surface. The saliva evaporates, taking along heat. Since the dog loses heat, it cools off.

Chill Out

You're covered. On top of your skin sits a thin layer of air. When you're not moving, this body-hugging air space stays in place, surrounding every part of your exposed body.

Your personal air covering interferes with the free flow of heat. It is an invisible barrier that cuts down on the amount of energy that leaves your body. Since heat is kept in, you feel warmer.

Now imagine a wind whisking away this airy layer of insulation. Since your cover is blown away, heat flows more quickly to your skin's surface. At the surface, the blowing wind continues to "steal" heat, making you feel colder.

WIND CHILL

The loss of body heat by the wind is called *wind chill*. As you might have guessed, the faster the wind, the quicker the heat loss. The slower the wind, the slower the loss of heat.

Some people call the wind chill factor a "feels-like" number. That's because the factor tries to put a number on how a person feels as heat is taken away. Remember that wind chill temperature isn't the actual outside temperature. It is, however, an ideal way of warning you to "bundle up" because it's going to feel way colder than the thermometer reading!

SAMPLE CHILL

If it is 0°F and the wind is blowing at 15 mph (24 km/hr), the wind chill can make it feel as if it's nearly −20°F! With this chill, it takes less than 15 minutes for frostbite to occur.

NEW EQUATION, OLD FACTOR

Things change. Even scientists rethink their formulas. The original formula for figuring out wind chill was developed in the mid-1900s. It was based upon experiments done by explorers in Antarctica. These scientists observed how different wind speeds and air temperatures affected the rate at which a can of water froze solid.

In the fall of 2001, a new wind chill formula was unveiled. This one was based upon a better understanding of how heat transfers. It also took into consideration human subjects who were exposed to blasts of cold air within a wind tunnel. The actual temperature drop on the subjects' skin was used to help make up the new equation.

WIND TUNNEL RESEARCH

Bad Hair Day

Ever have a bad hair day? No matter what you do, your mop won't stay in place. On humid days, it's guaranteed to pop, frizz, and spring out of control.

Bad hair days have a meteorological connection. It's the humidity in the air that produces the uncontrolled *boing*. Consider this: A strand of hair that's about a foot long can vary about ½ inch (1.3 cm) in length depending upon the humidity of the air. This stretching property of hair was put to use in the first hygrometer back in 1783. It was so reliable that hair hygrometers were used by weather stations into the 1960s. They were eventually replaced by more reliable electric hygrometers.

HAIR HYGROMETER

You'll Need

A milk carton	Toothpick
Scissors	A long strand of hair
Pin	Two damp paper towels
Tape	Penny

To Do

1. Have an adult cut an "H" pattern near one end of the carton. Bend up the cuts into two small flaps. Have an adult push a pin through the center of both flaps.

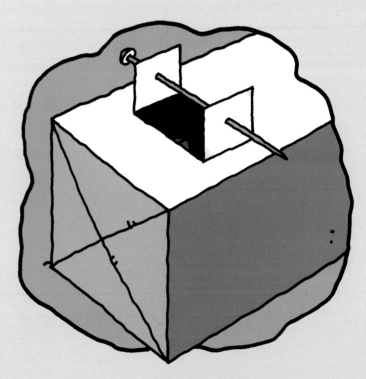

2. Stick a toothpick on the pin's point. Beneath the toothpick, draw a scale. Although the scale won't measure specific values, it can be used to compare different levels of humidity. So feel free to make up your own numbers.

3. Place a strand of hair between two damp paper towels. Let the hair remain in this high-humidity sandwich for 10 minutes.

4. Tape one end of a strand to the far end of the container. Wrap the hair several times around the pin and then attach this end of the strand to a penny. Let the penny hang off the side of the container. Its weight should pull the hair tight.

5. Position the toothpick so that it points to 100% humidity. Over time, the hair will dry out and shrink. As it does this, it will turn the pin and the toothpick will show lower humidity levels.

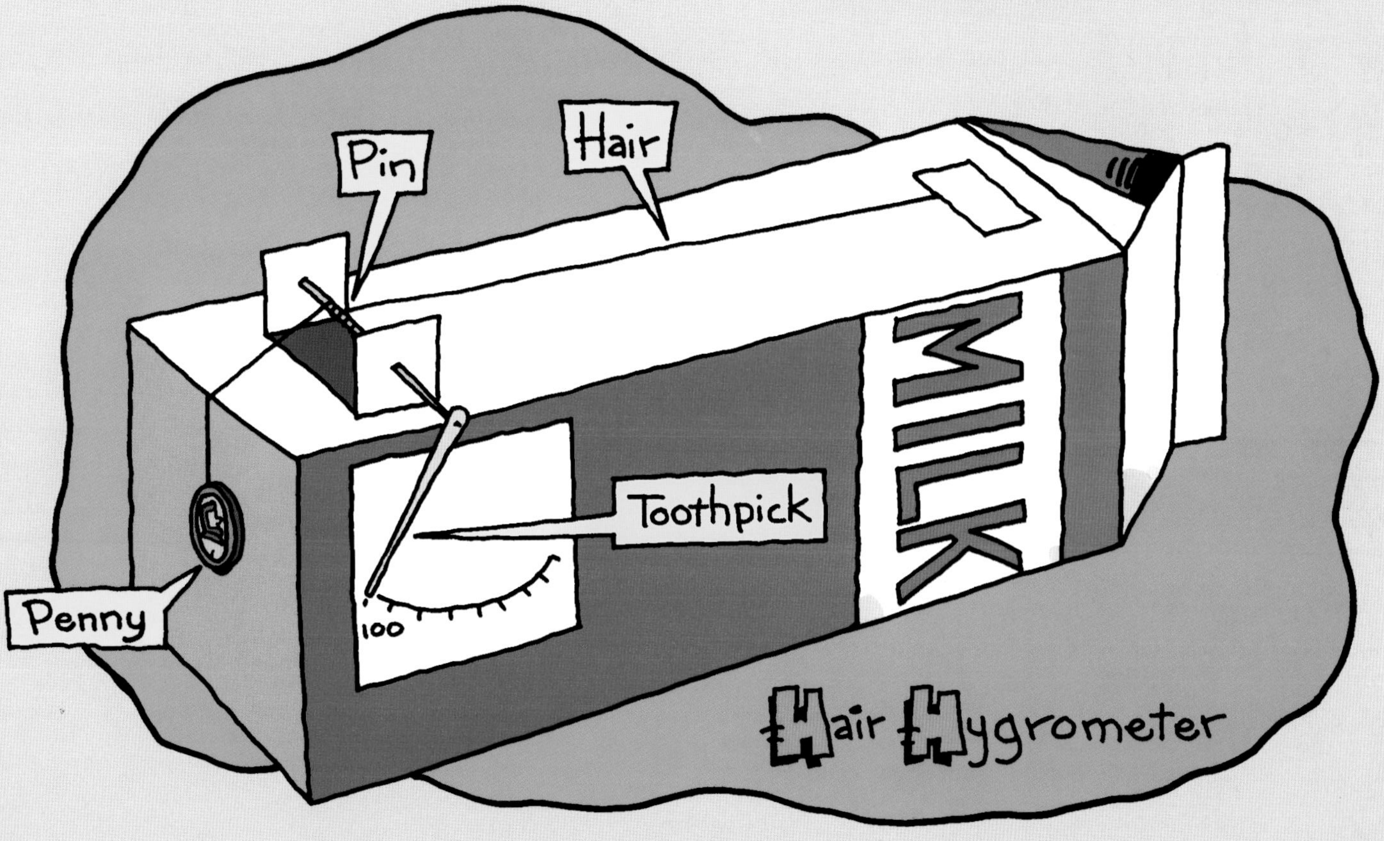

Pin

Hair

Toothpick

MILK

Penny

100

Hair Hygrometer

The Wet and Dry of It

Here's another weather instrument that measures relative humidity. As you'll see, the operation of this tool is based upon how quickly air can take in additional moisture. Air that is already full of moisture won't easily absorb more water. In contrast, air that is dry has space to hold water vapor.

Consider a dry thermometer. Since the thermometer is not "wearing" a wet jacket, it cannot be cooled by evaporation. Place this dry thermometer in the air and it measures the actual air temperature—not the temperature that is connected to humidity.

Now consider a "wet" thermometer. Its bulb is wrapped in a sleeve that is soaked with water. Unlike its dry cousin, the wet bulb is influenced by evaporation. When water evaporates from the surrounding sleeve, it takes away heat. This loss of heat drops the temperature reading of the wet bulb.

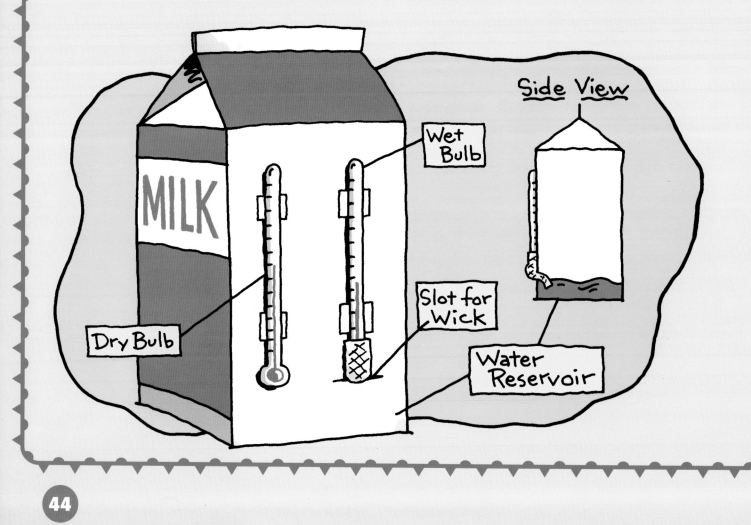

MILK

Dry Bulb

Wet Bulb

Slot for Wick

Side View

Water Reservoir

FINDING RELATIVE HUMIDITY

You'll Need

A milk carton
Scissors
Tape
Two thermometers
A piece of muslin fabric
   (such as a wick for an oil lamp)
Water

To Do

1. Have an adult cut a slot into the milk carton about two-thirds down the carton's length.
2. Use tape to attach two thermometers to the outside of the milk carton as shown in the illustration.
3. Place the muslin sleeve over one of the thermometers. Insert the other end of the sleeve through the slot into the milk carton.
4. Add water to the container. Make sure that the level remains below the slot but that it keeps the muslin constantly wet.
5. Read the temperatures from both the dry-bulb and wet-bulb thermometer. Then, subtract the wet-bulb temperature from the dry-bulb temperature. Use this difference and the dry-bulb temperature to find the relative humidity on this humidity chart.

RELATIVE HUMIDITY TABLE (shown in %)

Dry bulb	Difference between wet- and dry-bulb readings in °F			
°F	1	2	3	4
60	90	80	71	61
64	91	82	73	65
68	91	83	74	66
72	92	83	76	68
76	92	84	77	69
80	92	85	78	71

Dew Tell

Hot air can hold a great deal of moisture. That's because heated air molecules move quickly. Fast-moving molecules spread out and create spaces and gaps that can be filled by molecules of water.

Now imagine if we begin cooling a volume of warm air. As the molecules lose heat, they slow down. At slower speeds, the molecules begin to come together. As they do this, the spaces between neighboring air particles shrink. For a molecule of water vapor, this means a whole bunch of "No Vacancy" signs are springing up.

As the temperature continues to drop, less space is available. Eventually, you reach a temperature at which the air is holding its maximum amount of water vapor. This temperature is called the *dew point*. If the temperature falls below the dew point, some of the vapor will get "kicked out" of the gas phase. When this happens, the vapor condenses into a liquid.

THINK QUICK

The dew drops that collect on blades of grass come from:

a) Leaky sprinkler systems.
b) Animals that drink too much.
c) Water vapor in the nighttime air.

Answer: c) Water vapor in the nighttime air. In the early evening, the air is warm and humid. Once the sun has set, the evening air cools. The grass cools quickest and its temperature falls below the dew point. Water vapor condenses on the plant surface to provide the droplets of water we call dew.

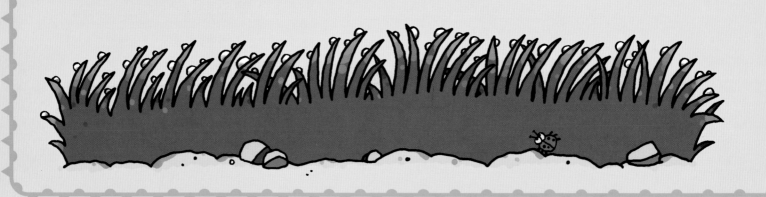

DEW THIS

We'll uncover the dew point in this activity.

You'll Need

Soup or coffee can
Warm water
Spoon
Crushed ice
Outdoor thermometer

To Do

1. Fill the can with warm water. Add a spoonful of crushed ice. Stir the mixture with the spoon.

2. Use the thermometer to monitor the temperature of the water.

> **CAUTION:**
> Thermometers are fragile. Use only an outdoor thermometer protected by a plastic or metal sleeve. Do not use the thermometer to stir the ice-water mixture.

3. Observe the sides of the can for droplets of condensation. If you don't see any droplets, then keep adding more ice until they appear. Be sure to use only the spoon to stir the mixture. At the first sign of droplets, record the temperature as the dew point of the surrounding air.

Cloud Types

Imagine a mass of humid air near the ground. Suppose the air begins to heat up. As it warms, its density decreases. Since it becomes lighter, the air mass rises. As it goes up, the air travels into cooler surroundings. The cooler temperatures condense the rising water vapor. The gas changes into a liquid and forms tiny droplets on particles of dust or other surfaces. These tiny droplets form the fluff in clouds. As more water vapor condenses, the droplets grow in size. Soon the droplets grow to about one million times their original size. At this much larger size, the droplets are too heavy to be kept up by air currents. They fall to the ground as rain.

BACK TO BASICS

There are three basic types of clouds: cumulus clouds, stratus clouds, and cirrus clouds.

Cumulus (CUE-muhl-luz) clouds have a fluffy appearance that may resemble puffs of cotton. Most cumulus clouds are found in the lower half of the troposphere. Rain clouds (cumulonimbus) can form towering structures that can stretch skyward through the entire troposphere!

Stratus (STRAH-tus) clouds have a layered appearance. They are often found close to the ground. Nimbostratus clouds are low-lying, gray clouds. Although they seldom produce intense showers, they often keep things wet with their damp, misty drizzle.

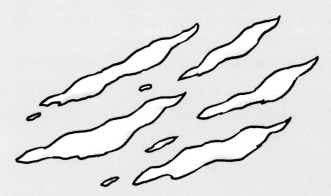

Cirrus (SEAR-rus) clouds have a feathery appearance. They are the highest of the cloud types and form at the upper edge of the troposphere. At these high altitudes, all of the cloud droplets are frozen into ice crystals.

FOG

Fog is a type of cloud that forms on or near the ground. One type of fog forms at night when ground temperatures drop below the dew point. Another type of fog forms where cooler land meets warmer water. Fog can also form in places where water currents of different temperatures clash. In the North Atlantic, this type of fog can hide dangerous icebergs found within the cold water flow.

CLOUD MAKING

Add an inch of warm water to a wide mouth, clear plastic jar. Place a strainer in the mouth of the jar. Add ice cubes to the strainer. Watch what happens when the air cooled by the ice meets the humid and warm air rising from the container bottom.

Ice →

Warm ↗
Water

Facing Fronts

Imagine the air that "sits" on top of the freezing midsection of Canada. Like the ground below, the air becomes chilled. In fact, during winter, this mass of air is downright *cold*.

Now consider another mass of air that forms over the sun-baked land of Mexico. Heat transfers from the land to the atmosphere, warming this overlying blanket of air. This mass of air is *warm*.

As you know, air masses don't stand still. They move. And as they move, they bring weather with them. The leading edge of an air mass is called a *front*. So when that Canadian air mass moves south, it's the cold front that announces the upcoming drop in temperature. Likewise, a warm front is an introduction to higher temperatures. But fronts do much more than announce a change in temperature. They make weather.

COLD FRONT: RAPID CLIMB

Since cold air is heavier than warm air, it stays near the ground. So as the cold front moves into a warmer region, it rapidly lifts the surrounding warm air. In ways, it acts as a steep wedge, sending warm and moist air skyward.

As the warm, humid air is pushed upward, it cools. This quick cooling produces condensation and precipitation. This is the ideal setting for the formation of thunderstorms. The upward push creates rapid and violent air currents. The quick cooling condenses water vapor and produces intense, but mostly short-lived cloud bursts.

Thunderheads →

COLD FRONT

WARM AIR

WARM FRONT:
SLOW AND STEADY

Now imagine an approaching warm front. As this air mass moves in, it pushes the cold front ahead. However, at the same time, it also rises above the cold air that is already in place.

The climb above the cold air is slow and steady. It doesn't cause any strong upward push of air and sudden rainstorms. Instead, it produces a gradual thickening of clouds. These clouds can stretch hundreds of miles before it rains or snows. When it finally happens, the rain or snow is steady and long-lasting.

STATIONARY FRONT:
GOING NOWHERE

As its name suggests, a stationary front doesn't move. Mostly it stays in place. If this stationary front was producing rain, then the rain might continue to fall until the front moves on. This type of condition might lead to flooding.

OCCLUDED FRONT

An occluded (OH-clue-did) front has a mass of air that gets carried up. The most common occluded front occurs during a winter storm. Warm air gets "occluded" as it rises above the boundary between a cool and cold air mass.

Going Down

Once evaporated from the Earth's surface, how long will the average molecule of water remain in the atmosphere?

a) One day.
b) One week.
c) One year.

Answer: b) One week. Once a water molecule evaporates, it rises and enters the air above. Within the atmosphere, the average particle remains up for about a week to 10 days. Eventually, the water vapor will cool and condense back into liquid water. As these water droplets grow, they become too heavy to remain in the clouds. When this happens, the droplets fall from the sky as precipitation.

WHAT'S SNOW?

Ice crystals that fall through clouds arrange themselves into six-sided shapes we call snowflakes. These intricate designs come from the tiny pushes and pulls of hydrogen atoms. Since the pushes and pulls can be arranged in all sorts of patterns, snowflakes don't look alike.

WHAT'S HAIL?

Hail is large, frozen raindrops. As liquid raindrops fall through a layer of freezing air, they become solid. Upward wind currents take the frozen raindrops back up to the cloud top where more water collects on their surface. Again, they fall and another coating freezes onto the hailstone. As the process continues, the hailstone grows in size until it is too heavy to be kept up by the winds.

WHAT'S SLEET?

Sleet consists of tiny ice pellets that seem to accompany rain or snow showers. Sleet starts as falling snow that drops through a warm layer of air. The flakes partially melt and refreeze as they continue to fall. Unlike rain that may freeze to surfaces, sleet doesn't coat the ground or accumulate.

Often, you can see it bouncing on the ground or car windshield before it soon melts away.

WHAT'S SNU?

Nothing much. What's snu with you?

Strange Rain

On July 12, 1873, a rainstorm that fell at Kansas City, Missouri, dropped a shower of:

a) Cats and dogs.
b) Frogs.
c) Really big raindrops.

Answer: b) Frogs. That's right, frogs—those slimy little amphibians that mostly live in streams and lakes. No longer were these frogs swimming in a watery world. Instead, they were falling from above. According to observers, there were so many frogs that their falling bodies darkened the sky.

WHAZ UP?

Did frogs really fall from the sky? It seems so. There are several recorded animal rainstorms, including showers of frogs and fish. Perhaps the grossest of all animal downpours was a drenching of maggots.

Falling to Earth is easy to explain. Gravity is the cause of that. Getting up into the sky is a totally different challenge.

Scientists believe that tornadoes and smaller whirlwinds are responsible for the flying ability of these animals. As these winds rip around, they can suck up what's in small ponds and lakes. So along with a good deal of water, up goes the pond life.

But whatever goes up must come down. As the winds die out, the animals carried up will fall back to Earth.

Support for the "twister theory" comes from a 1995 event in which a twister picked up soft drink cans from a bottling plant. The cans were carried up by the tornado's strong winds. About 150 miles (241 km) from where the factory was located, the cans fell as a downpour.

HAIL TOO

Falling animals aren't limited to rainstorms. During a hailstorm, a turtle that was completely frozen in a block of ice plummeted to the ground. Not only had this animal been lifted up by a tornado's winds, but its unfortunate journey took it through wet, freezing air.

Charge It!

FLASHY BEGINNING

Blow up a balloon. Stroke the inflated balloon with a piece of wool (from a sweater, scarf, or hat). Get a fluorescent lightbulb from an adult. Darken the room. Have the adult hold the bulb. Slowly bring the charged balloon to the bulb's metal pins. As a spark jumps from balloon to the metal, the lightbulb flashes! The flash was produced by the tiny "bolt" of electricity that jumped through the air.

UP IN THE AIR

Within a cloud, particles of air, dust, and water vapor crash together. Like stroking a balloon with a piece of wool, the colliding of particles causes the buildup of static electricity. The charges that separate within the cloud can jump to the ground below. We call this giant discharge lightning!

FOLLOW THE LEADER

Lightning begins when some of the negative charges move from the cloud to the ground. This movement starts things off. That's why it's called the "leader." Not only does the leader set the path for the lightning bolt, but it makes the surrounding air electrically unstable. Charges become free to move.

A split second later, a huge amount of electricity follows the leader's path. Electrons in the unstable air drain into the ground below. This rapid and major release of electricity from the air to the ground produces a huge flash of lightning. Since the flash moves upward, it is called a return stroke!

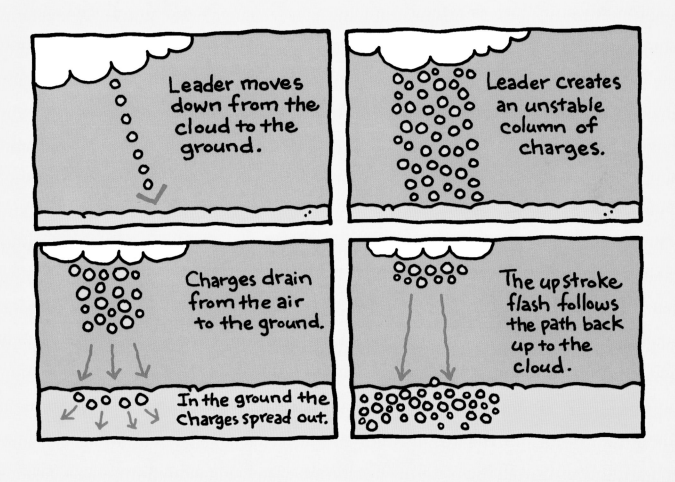

Leader moves down from the cloud to the ground.

Leader creates an unstable column of charges.

Charges drain from the air to the ground.

In the ground the charges spread out.

The upstroke flash follows the path back up to the cloud.

Lightning and Thunder

True or False?

Ben Franklin held the string of a kite that was struck by lightning. He used this shocking experience to help create a theory of electricity.

False. Don't believe everything you read in books. Had lightning actually struck the kite he was holding onto, Franklin would have been fried. Most likely it was the buildup of static electricity on the kite and string that created any sparks that he saw.

Hot Stuff

The lightning bolt heats the surrounding air to five times the temperature of the sun's surface. This causes the air to violently expand and produce the sonic boom we know as thunder.

COUNT IT OUT

The thunder caused by a lightning bolt travels much slower than the lightning bolt's flash. You can use this difference to calculate your distance to a lightning flash. Here's how it works:

Light is so fast that we'll assume that we see the flash at the exact moment that the lightning strikes. It's the speed of sound that we'll use to estimate the distance to the strike. Sound travels at about one-fifth of a mile (1000 feet or 305 meters) per second. So if it takes 10 seconds to hear thunder after seeing the flash, you know that the lightning is about 2 miles away. And don't forget, this timing trick also works with estimating the distance to fireworks!

Impress an Adult!

The most common lightning doesn't travel from the cloud to ground. Most often, lightning bolts travel from cloud to cloud. Since they are on top of the low-lying clouds, you can't see them. That's why you can sometimes hear thunder, but not observe the lightning flash.

Think Safety

When lightning is nearby, take cover in a building. Don't stand beneath a tree or other tall objects. Cars are usually safe to be in, since the car body carries the electricity directly to the ground. But above all, don't fly kites!

Tornadoes

Imagine a piece of straw going through a telephone pole. Sounds unbelievable? It's not. The incredible winds of a tornado can propel all sorts of objects at speeds greater than 250 mph (402 km/hr). At that speed, even a piece of straw becomes a deadly missile!

A TWISTED TALE

Tornadoes are powerful weather events. They begin as violent winds within a thunder cloud. As the tornado develops, its winds twist and turn sideways. The winds soon form a funnel that moves downward towards the ground. When the funnel "touches down," its whirlwinds can rip a path of destruction.

The funnel, or vortex (VOR-teks), acts like a giant vacuum cleaner. The winds racing around the base of the vortex suck everything up. Most tornadoes last for only about 15 minutes. During these violent moments, objects such as houses, cars, and farm animals are sucked into this whirlwind and carried away.

Impress an Adult!

A tornado that forms over water is called a waterspout. Most waterspouts have been recorded in waters off the Florida coast where they can whip up winds to about 90 mph (150 km/hr). Some have even been observed in Lake Tahoe, a body of water high in the Sierra Nevada mountain range.

Deadliest Tornado

The deadliest single tornado to strike the United States occurred on March 18, 1925. It lasted for over 3 hours and tore up a path that was over 200 miles (322 km) long through Missouri, Illinois, and Indiana. When the tornado died out, nearly seven hundred people had lost their lives.

VISIBLE VORTEX

Do you want to see a vortex in action? If so, get two large plastic soft drink containers. Fill one with water. You can add food coloring for a more dramatic appearance. Tape the openings of both containers together with duct tape or some waterproof (and tough) material. Hold the bottles over the sink, then turn the container with the water upside down. Watch the water spiral down into the empty container.

Hurricanes

Where did the deadliest hurricane strike?

a) Galveston, Texas.
b) Islands of the West Indies.
c) Bangladesh.

a) Good try, but the Galveston hurricane was the deadliest super storm to strike the United States. An estimated 6000 to 8000 people died when this storm came ashore on the Texas coast on September 8, 1900. Since there were no warnings, the people were totally unprepared for this disaster. Offshore, the hurricane's raging winds produced a sudden 20-foot increase in the tide (called a storm surge) that flooded this port city.

b) Another good try. From October 10–16, 1780, a superstorm known as the "Great Hurricane of 1780" raged in the Caribbean. This hurricane killed over 20,000 people as it devastated the islands of Barbados, Martinique, and St. Eustatius. This superstorm holds the record for loss of life in the Western Hemisphere.

c) Correct. In November 1970, the world's deadliest weather disaster struck the country of Bangladesh. Check it out on a map and you'll see that Bangladesh is a low-lying Asian country located on the coast of the Indian Ocean. A huge storm surge flooded the heavily populated coastal regions killing between 300,000 to 500,000 people!

MOST AWESOME STORMS

Hurricanes are the most awesome storms on Earth! They are known in the Pacific as typhoons and by meteorologists as tropical cyclones. The largest of these storms can stretch up to 500 miles (800 km) in diameter.

What makes a hurricane different from other storms? Just peek under their thick cloud blanket (using radar, of course) and you'll discover that hurricanes are made up of bands of thunderstorms. Like water going down a drain, these storm bands spiral inward towards the center of an organized structure. The center of this structure is called the *eye*. Unlike the rest of the hurricane in which winds may gust to almost 240 mph (386 kmph), the eye remains mostly calm. But there *is* moving air in the eye. It's a slow-moving column of air that pushes downward and squishes weather and winds. That's why the eye is calm!

THINK QUICK!

The energy that powers a hurricane comes from:

a) Tidal waves.
b) Warm ocean water.
c) A very large wall outlet.

Answer: b) Warm ocean water. The heat energy that is in warm ocean water powers a hurricane. This energy is transferred from the ocean to the air that lies above it and churns the atmosphere. These violent air movements produce thunderstorms.

As they continue to absorb energy, the storms grow. Soon neighboring storms become organized and join into a large swirl. At the center of this swirl is a region of low pressure called a tropical depression. As the depression grows stronger, it evolves into the "eye" of the storm. When the speed of the surrounding winds reaches 74 mph (119 km/hr), the tropical storm earns hurricane status.

WATCHES AND WARNINGS

A hurricane watch is issued when it appears that hurricane winds will arrive within 24 to 36 hours. As the storm gets closer, the watch may turn to a warning. When the hurricane winds are expected in less than 24 hours, a hurricane warning is issued.

DEATH OF A HURRICANE

As the hurricane moves over land, it breaks contact with the ocean water that was supplying its energy. Eventually, it weakens and its structure falls apart.

HURRICANE CATEGORIES

Hurricanes can be classified by wind speed and damage potential into five categories. Here's how those categories stack up.

Category	Wind speed	Storm surge*	Damage potential
Category 1	74–95 mph (119–153 km/hr)	4–5 feet (1–1.5 m)	Minimal
Category 2	96–110 mph (154–177 km/hr)	6–8 feet (2–2.4 m)	Moderate
Category 3	111–130 mph (179–209 km/hr)	9–12 feet (3–3.7 m)	Extensive
Category 4	131–155 mph (211–249 km/hr)	13–18 feet (4–5.5 m)	Extreme
Category 5	More than 155 mph (249 km/hr)	More than 18 feet (+5.5 m)	Catastrophic

* A storm surge is a quick rise in water level that is produced by the approaching hurricane. It is not a huge "wall of water" or giant wave. The surge is a fast and steady rise of the tide.

Weather Maps

What's the best way to predict the future?

a) Buy a crate of crystal balls.
b) Install a DSL line to a psychic advisor.
c) Learn how to read a weather map.

As you probably guessed (after all this *isn't* a book about the paranormal), your best gateway to the future is learning how to read a weather map. A weather map not only tells the current conditions, but it can be used to make accurate weather predictions.

Some weather maps show only the temperature and the type of weather that is happening. As you might imagine, it's hard to use this limited information to predict the future. To get a better idea of what is happening (and what is going to happen), you need to know about fronts, winds, and pressure systems. You need to know where things are going!

WHY BOTHER WITH HIGHS AND LOWS?

The uppercase "H" that is drawn on this weather map points out the center of a *high pressure* region. Likewise, the uppercase "L" stands for a *low pressure* region. Big deal, you might think. So one region has a slightly different pressure than the surrounding air. Why should I care? Shouldn't I be more concerned with approaching warm and cold fronts?

Knowing the pressure of an air mass tells you plenty about what type of weather to expect. A low pressure region is associated with unstable air and stormy conditions. In contrast, a high pressure system sets the scene for mostly clear, cloudless weather.

FRONT PRIMER

Remember the four fronts we discussed in Facing Fronts (see pages 50–51)? Well, here's how to illustrate them on the weather map. The moving fronts move in the direction of the triangles and hills.

Cold Front

Warm Front

Occluded Front

Stationary Front

Read Between the Lines

Have you ever seen a weather map covered with a pattern of lines? Those weather map lines are called *isobars* (EYE-so-bars). Isobars are most often included on the maps that forecasters use to predict weather. As you may have guessed (from the *bar* in isobar), these lines connect places of equal pressure.

Since winds are produced by changes in pressure, you can use isobars to predict wind speed and direction. For example, when isobars are spread out, there is a gradual change in air pressure. Since the change is not abrupt, the air remains mostly calm.

When the isobars are packed closely together, there is a rapid change in pressure. Quick changes produce strong winds as the force moves from the intense high pressure region to the nearby region of low pressure.

WIND FLAGS

In addition to showing isobars, weather forecaster maps have really cool symbols. The basic part of the symbol is called a *wind flag*. The flag identifies wind direction, speed, and sky cover. Additional information, such

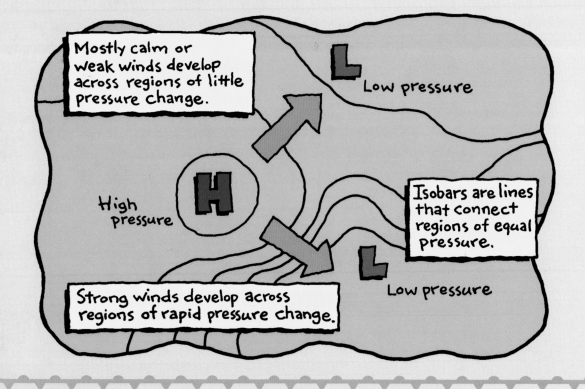

Mostly calm or weak winds develop across regions of little pressure change.

Low pressure

High pressure

Isobars are lines that connect regions of equal pressure.

Strong winds develop across regions of rapid pressure change.

Low pressure

as temperature, dew point, current weather, and chance of precipitation are often included next to the symbol.

The central part of the wind flag is a circle. If the sky is clear, the circle is not filled in. However, as more clouds cover the sky, the circle becomes more "penciled in" like the answer blocks on a standardized test. Here are a few of the ways that the circle can appear:

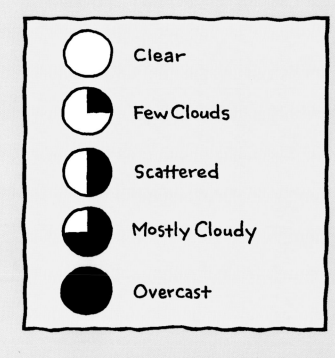

A line that extends from the circle points to the direction from which the wind is coming. By "reading" the circle and line, you can tell the wind direction and sky cover. See what we mean?

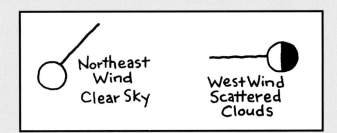

When there's no wind, the sky circle is drawn as a double circle. A 1–2 mph wind is shown by a plain line. Barbs that "hang" from the line show various wind speeds.

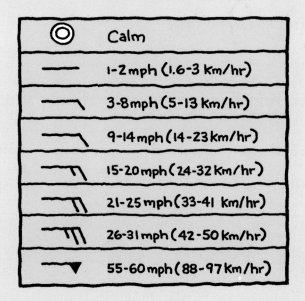

Eyes in the Sky

How far away are the orbits of most weather satellites?

a) About the same distance as an orbiting shuttle.

b) About twice the distance of an orbiting shuttle.

c) Over 150 times the distance of an orbiting shuttle.

Answer: c) As incredible as it may seem, those weather satellites circle our planet at an altitude over 150 times the distance of a low-orbit shuttle mission. In other words, the satellites are way, way out there. The path followed by these satellites is called a *geostationary orbit*. This special orbit is located at a distance of 22,238 miles (about 36,000 km) above the equator.

GOES Figure

The fleet of weather satellites that remain in fixed orbit above the Earth are the eyes of the GOES (short for Geostationary Operational Environmental Satellites) project.

Within a geostationary orbit, the satellite moves at the same speed as the Earth's spin. This keeps the satellite fixed over the same spot of land. Since it stays above the same region, the satellite keeps a fixed view of Earth. This is critical to watching weather systems move across the planet's surface.

Currently, we have five weather satellites that are locked in a geostationary orbit. The five are spread out so that a complete view of the Earth's surface is constantly monitored. One of the satellites is parked above the eastern end of the Pacific Ocean. Another remains stationary over the western end of the Atlantic. With cameras aimed at the Earth, these two satellites give us a full view of the North American continent.

The weather satellites are somewhat limited in what they can do. They are mostly used to capture photographs that show the cloud covering of Earth. It is these "tweaked" images that appear on the nightly weather reports. These satellites can also look through the atmosphere and capture measurements of ground temperatures.

There are two other satellites dedicated to weather watching. These craft continually circle the planet on an orbit that takes them around both the North and South Poles. Since they are not geostationary, they have a much lower orbit of just over 500 miles (805 km).

Proverbial Sky Watching

Red sky at night, sailor's delight.
*Red sky in morning, sailors take warning.**

** For the lowdown about this proverb,
see the Real Story below.*

What's the weather going to be?

Do you know any good proverbs?

Not another proverb! Ever wonder who makes up these tired sayings? And besides, are they even correct?

Most weather proverbs were made up years ago as a method of communicating ways to predict weather. Based upon observations, most of these teachings were recorded as simple rhymes and sayings. Nowadays, modern forecasting tools have replaced our need to rely on these proverbs.

THE REAL STORY

The "Red Sky" proverb is meant to work for someone who is in a place where the weather moves from west to east. If the western sky is clear of humidity, then sunlight from the setting sun will be scattered by dry dust. This scattering separates colors and produces a red sky. Hence, the weather that will arrive from the west should be clear. If the eastern sky in which the sun rises is red, then the dry air has passed and the scene may be set for wetter weather to arrive from the west.

MIX AND MATCH

Can you match each of the three proverbs with its weather event?

1. *When stars shine clear and bright,*
   We will have a very cold night.

2. *Rain long foretold, long last,*
   Short notice, soon will pass.

3. *A rainbow in the morning is the shepherd's warning.*
   A rainbow at night is the shepherd's delight.

a) Rainbows are produced by water in the atmosphere. To see one you need to face away from the sun. Therefore, a morning rainbow means that wet weather is due to arrive from the west. Similarly, a rainbow in the evening means that the humid weather has moved to the east.

b) Clouds reflect heat back to the ground. On a clear night, the temperature drops sharply because of a quick cooling of the Earth's surface.

c) A warm front that brings steady precipitation produces a thick cloud covering many hours before the rain. In contrast, an arriving cold front produces fast-moving and sudden thunderstorms.

ANSWER
1. b; 2. c; 3. a.

Telling Tails (and Other Natural Signs)

For thousands of years, people searched for ways to predict the weather. Since tools like thermometers, anemometers, and weather satellites wouldn't be invented for some time, they had to rely on things at hand.

Check out the natural predictors of weather below. Which ones seem to work? Which ones seem to be mostly a tall tale? Make a guess, then check out the true story.

SQUIRREL TAILS

It was said that a squirrel's appearance and behavior can be used to predict weather. If the squirrel has a bushy tail during the fall season, it's sure to be a cold winter. You could also check out the squirrel's store of nuts! If the storage is large, then a long cold winter will follow.

PINE CONES

The scales on a pine cone can predict rain. When the scales are spread apart and stand out, the weather will be dry. If the scales are packed closely together, then there is a good chance that rain will soon follow.

WOOLY BEAR CATERPILLAR

The width of this caterpillar's brown banding indicates the severity of the upcoming winter. If the brown band is wide, then a mild winter is expected. If the band is narrow, then a cold winter is on its way.

LOW-FLYING SWALLOWS

When these birds fly close to the ground, expect bad weather.

GROUNDHOG

If this animal sees its shadow on February 2 (Groundhog Day), then the weather will remain cold for six more weeks. If it doesn't see its shadow, then expect an early spring.

ANSWERS

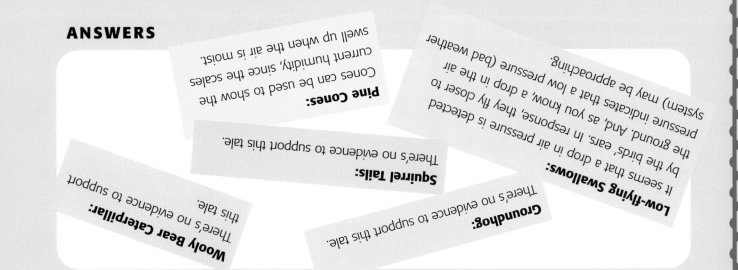

Pine Cones:
Cones can be used to show the current humidity, since the scales swell up when the air is moist.

Squirrel Tails:
There's no evidence to support this tale.

Low-flying Swallows:
It seems that a drop in air pressure is detected by the birds' ears. In response, they fly closer to the ground. And, as you know, a drop in the air pressure indicates that a low pressure (bad weather system) may be approaching.

Wooly Bear Caterpillar:
There's no evidence to support this tale.

Groundhog:
There's no evidence to support this tale.

A NATURAL THERMOMETER

Listen to the chirps of crickets. Then count the number of chirps they make during a 14-second period. Add forty to this number. Your final number will give you the approximate temperature in Fahrenheit degrees.

This formula works because crickets are cold-blooded animals and the rate at which they chirp depends upon the surrounding temperature. The higher the temperature, the greater the number of chirps.

Colors in the Sky

Why is the sky blue?

a) Because.

b) So that it matches the wall paint with the same name.

c) It's a scattering of light thing. Sunlight contains all the colors in the rainbow. When this white light strikes the atmosphere, blue scatters most* as it bumps into oxygen particles. This scattering sends the color blue in all directions. So no matter where in the sky you look, you see the blue light that is being scattered down by that region of the sky. That's why the sky appears blue.

Answer: c) With an explanation that long, how couldn't that be the correct answer? And besides, we'll throw in this illustration to further show our point.

*Actually, violet (the shortest wavelength) scatters most, but your eyes are more sensitive to blue light.

RAINBOWS

Did you know that light bouncing around *inside* of a raindrop produces a rainbow? For you scientific thinkers, here's what happens:

1. White light (containing all colors) strikes the near side of a raindrop and enters this bead of water.
2. The light travels through the raindrop and hits the inner surface of the far side of the drop. This surface acts like a mirror and reflects the light.
3. The reflected light exits back out the side through which it first entered.
4. Things, however, have changed. As the light moves both times through the boundary of air and water, its path gets bent. Since different colors bend at different angles, the white light gets spread out into a rainbow of colors.

So you don't think we're playing favorites to blue sky people, here's a drawing that shows the formation of a rainbow.

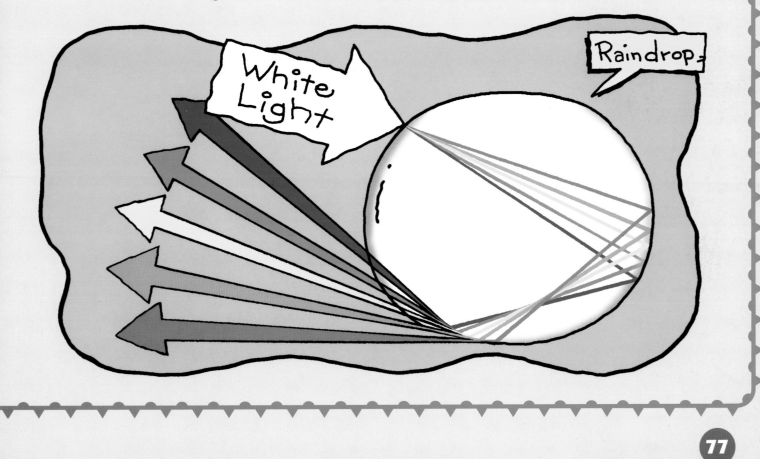

The Future

What's the future weather going to be?

a) Get ready for a whole lot of warming.

b) Look out for a little ice age.

c) No one knows for certain.

Answer: c) No one knows for certain. However, temperatures seem to be warming up (and the winters becoming less extreme). Check out high temperatures and you'll find that most record breakers occurred in very recent years! In fact, if you do the math, you'll uncover that the warmest decade on record was the 1990s!

What does this mean? Are we locked into a warming trend? Should you invest in central air conditioning? Or is this warm-up an introduction to an upcoming ice age? Will swimmers soon be trading in their trunks for ice skates? Again, no one knows for certain.

Temperatures have gone up and down throughout history. They seem to be part of the natural cycle. Plus, we seem to be getting over a recent cold spell that occurred around 1450–1700. During that time, people most likely experienced the coldest year (1601) and the coldest century (17th). Brrrrrrrrrrrrrrr.

GREENHOUSE EFFECT

Have you ever entered a car that has been "baking" in the summer's heat? You know the setting: Windows rolled up, no shade, and bright sunlight pouring in through the windows. If so, you know *not* to touch any metal or sit on the seats with bare skin. They are hot, hot, hot! In fact, the whole inside of the car feels like an oven.

The heating of the car's inside is a result of something called the *greenhouse effect*. The greenhouse effect begins as light energy passes through transparent materials, such as glass or atmospheric gases. Beneath this covering, the light continues until it eventually strikes a surface. When it makes contact, some of the light energy is released as heat. Unlike the light that traveled through the transparent cover, this heat energy can't get out. It remains trapped and continues bouncing around and releasing more heat. The surroundings absorb the energy and rise in temperature.

GREENHOUSE GASES

Many scientists believe that we are increasing the natural greenhouse effect. Human activities such as burning gasoline, oil, and coal release greenhouse gases. This extra gas increases the way heat gets trapped in the atmosphere. Since more of the incoming solar energy is kept in, the temperature rises even more. Some scientists fear that these added gases are responsible for our current warming trend. But, again, no one knows for sure.

About the Author

Michael Anthony DiSpezio is a renaissance educator who teaches, writes, and conducts teacher workshops throughout the world. He is the author of *Critical Thinking Puzzles*, *Great Critical Thinking Puzzles*, *Challenging Critical Thinking Puzzles*, *Visual Thinking Puzzles*, *Awesome Experiments in Electricity and Magnetism*, *Awesome Experiments in Force and Motion*, *Awesome Experiments in Light and Sound*, *Optical Illusion Magic*, *Simple Optical Illusion Experiments with Everyday Materials*, *Eye-Popping Optical Illusions*, *Map Mania*, and *Dino Mania* (all from Sterling). He is also the co-author of over two dozen elementary, middle, and high school science textbooks and has been a "hired creative-gun" for clients including The Weather Channel and Children's Television Workshop. He also develops activities for the classroom guides to *Discover* magazine and *Scientific American Frontiers*.

Michael was a contributor to the National Science Teachers Association's Pathways to Science Standards. This document set offers guidelines for moving the national science standards from vision to practice. Michael's work with the NSTA has also included authoring the critically acclaimed NSTA curriculum, *The Science of HIV*. These days, Michael is the curriculum architect for the JASON Academy, an on-line university that offers professional development courses for science teachers. To learn more about this topic and Michael's cool science activities, log on to www.Awesomescience.org.

Index